The Role of AI in the Oil Industry

Innovation, Sustainability, and the Global Energy Future

Prologue: The Revolution of the Oil Industry in the Age of Artificial Intelligence

The oil industry, the backbone of global industrial development for over a century, is facing an unprecedented shift. With a central role in the world economy and a direct impact on the lives of billions of people, this industry has overcome operational and geopolitical challenges throughout the decades. However, at the dawn of the 21st century, its most formidable challenge is the need to reinvent itself in a world driven by technology, sustainability, and the demand for efficient, clean energy.

The emergence of artificial intelligence (AI) is not just a technological advancement but a paradigm shift that promises to transform every aspect of operations in oil fields. From exploration to production, AI introduces unprecedented levels of operational efficiency, precision, and sustainability. This document offers an in-depth and comprehensive view of how AI is revolutionizing the oil industry, providing companies with the tools they need to thrive in an increasingly competitive and regulated environment.

A New Horizon for the Oil Industry

The use of AI in the oil industry spans a wide range of applications: real-time monitoring, predictive maintenance, energy optimization, and integration with renewable energy. These technological advancements not only enable companies to maximize production and minimize costs but also position them as key players in the transition toward a more sustainable energy model.

Economic projections show that implementing AI can reduce operational costs by 15-25% and increase overall efficiency by 20%, results that could be a game-changer in an industry traditionally characterized by narrow margins and high exposure to oil price volatility.

The Role of AI in a Volatile and Demanding Market

The global energy market is inherently volatile. Fluctuations in oil prices, geopolitical tensions, and shifting consumer demands create an unpredictable environment. AI offers a powerful solution by enabling dynamic real-time adjustments. Intelligent systems can analyze demand patterns, geological and operational conditions, and environmental regulations, providing companies with a significant strategic advantage.

Moreover, AI facilitates faster and more accurate decision-making in critical situations, reducing response times and increasing operational safety. This is particularly relevant in complex scenarios, where speed and precision can mean the difference between operational success and crisis.

Transformation of the Workforce and Corporate Culture

The implementation of AI also signals a profound shift in the composition and skills required of the workforce in the oil industry. Traditional roles are being replaced by highly technical and strategic positions, such as automation engineers, data analysts, and advanced maintenance technicians. This transition requires investment not only in technologies but also in the continuous training of personnel.

Companies that invest in employee training will be better positioned to fully leverage AI's potential, ensuring successful and sustainable integration. Moreover, collaboration between humans and machines creates a work environment where human expertise and AI's analytical capabilities complement each other, guaranteeing optimized operations and well-informed strategic decisions.

Sustainability and Environmental Responsibility: A Global Imperative

The pressure to operate more sustainably has never been greater. Environmental regulations are becoming increasingly stringent, and consumers and investors demand responsible operations. AI plays a crucial role in environmental management by optimizing resource consumption, reducing emissions, and preventing environmental disasters such as spills or leaks.

For instance, integrating AI systems into critical operations has demonstrated a 10-20% reduction in operational emissions, directly contributing to global sustainability goals. This advancement not only enhances companies' reputations but also opens new opportunities in markets that prioritize sustainability.

Key Challenges and the Road Ahead

However, the transition to an AI-driven oil industry is not without challenges. Cybersecurity emerges as a critical concern, as digitalization increases vulnerability to cyberattacks. Protecting critical infrastructure and operational data requires a robust strategy that combines advanced technologies with strict security protocols.

Similarly, AI integration must be approached with strategic planning that considers long-term implications for corporate culture, operations, and the labor market. Resistance to change, inadequate technological infrastructure, and the need for clear regulations are some of the obstacles companies must overcome.

A Revolution Underway

This document not only highlights the benefits and challenges of AI in the oil industry but also provides a roadmap for its successful implementation. Through detailed analyses, case studies, and quantitative projections, it offers a comprehensive guide for companies to navigate this technological revolution.

An Intelligent and Sustainable Future

The adoption of AI marks the beginning of a new era for the oil industry. Companies that embrace this technology with strategic vision and commitment will not only enhance their profitability and operational efficiency but also contribute to a more sustainable and responsible energy model. This prologue invites the reader to explore how AI is redefining a critical industry, paving the way for a future where technology and sustainability coexist as fundamental pillars of global progress.

Table of Contents

Prologue

- The Revolution of the Oil Industry in the Age of Artificial Intelligence

Chapter 1: Introduction

- The Oil Industry in the 21st Century
- The Transformative Role of Artificial Intelligence

Chapter 2: Historical Perspectives

- From the Industrial Revolution to the Digital Age

Chapter 3: Key AI Applications in the Oil Industry

- Optimization of ESP and PCP Pumps with AI
- Tank and Separator Management with AI
- Real-Time Control and Automation

Chapter 4: Economic and Operational Impact of AI

- Increased Operational Efficiency
- Sustainability and Reduction of Environmental Footprint
- Adapting to Volatile Markets

Chapter 5: Future Innovations

- Emerging Technologies: Blockchain and IoT
- The Convergence of Technologies

Chapter 6: Redefining Workforce Roles and Competencies

- Workforce Optimization and Cost Reduction
- Continuous Training and Development
- Human-AI Collaboration

Chapter 7: Challenges and Risks of AI Implementation

- Cybersecurity in the Digital Age
- Cultural and Adoption Barriers

Chapter 8: Success Stories

- Chevron: Predictive Maintenance in Offshore Wells

- Shell: Supply Chain Optimization
- ExxonMobil: Sustainability Improvements
- BP: Data-Driven Exploration
- TotalEnergies: Refining Process Optimization

Chapter 9: Geopolitical Implications

- Effects on the Global Market: Uneven AI Adoption

Chapter 10: Future Projections

- The Future of the Oil Industry with AI
- Consequences for Countries and Companies Not Adopting AI

Chapter 11: General Conclusion

- Leading the Transformation of the Oil Industry with AI

Chapter 12: Key References

- Relevant Studies and Authors

Appendices

- Glossary of Technical Terms
- Detailed Case Studies

Introduction: The AI Revolution in the Oil Industry

The advancement of artificial intelligence (AI) marks a turning point in various industries, and the oil industry is no exception. For decades, oil fields have relied on the accumulated expertise of engineers, technicians, and operators, alongside specific technologies to maximize production and minimize risks. However, with the increasing complexity of operations, constant fluctuations in global markets, and the pressure to meet stricter environmental standards, the industry faces challenges that demand more advanced and adaptive solutions. In this context, AI has emerged as a fundamental tool for revolutionizing the management of oil fields.

The Transformative Role of AI

AI offers a range of innovative solutions addressing critical points in oil operations. From predictive monitoring of key equipment such as ESP (Electric Submersible Pump) and PCP (Progressive Cavity Pump) systems to real-time automation of complex systems like separators and tanks, AI enables constant oversight and proactive adjustment of operational conditions. These technologies not only improve operational efficiency and safety but also significantly reduce downtime, one of the industry's highest operational costs.

Key Benefits of AI Integration

- **Optimization of Human and Technical Resources:**
  AI automates routine and repetitive tasks, freeing human personnel to focus on strategic, high-value activities such as critical decision-making and operational planning.
- **Predictive Maintenance:**
  By analyzing real-time data, AI can predict equipment failures before they occur, enabling planned interventions that minimize disruptions and associated costs.
- **Energy Savings and Operational Cost Reduction:**
  AI algorithms continuously adjust operational parameters to maximize energy efficiency, translating into cost reductions and a smaller environmental impact.

Responding to Global Market Demands

The oil industry operates in a constantly volatile environment, influenced by geopolitical factors, crude price fluctuations, and changing demand. AI's ability to analyze large volumes of data and provide actionable insights in real time enables companies to quickly adapt to these shifting conditions. This ensures continuous and efficient production while positioning companies to compete effectively in an increasingly demanding global market.

Meeting Sustainability and Environmental Regulations

The pressure to reduce carbon emissions and comply with stricter environmental regulations is a growing challenge for the industry. AI directly contributes to these goals by optimizing processes to reduce waste, minimize energy consumption, and prevent incidents such as leaks and spills. Moreover, implementing AI solutions enhances operational transparency, facilitating regulatory compliance and improving corporate reputation.

A More Resilient and Sustainable Business Model

Integrating AI is not just a competitive advantage; it is a strategic necessity for ensuring long-term sustainability. In an environment where operational efficiency and environmental responsibility are increasingly critical, adopting advanced technologies enables oil companies not only to survive but to thrive.

Conclusion of the Introduction

In summary, the AI revolution in the oil industry represents a unique opportunity to completely transform operational management, improve efficiency, and reduce costs while addressing environmental and market challenges. Companies that embrace this technology with strategic foresight will be better positioned to lead in a constantly evolving global environment, ensuring their relevance and sustainability in the decades to come.

Chapter: Historical Perspectives - From the Industrial Revolution to the Digital Age

The history of the oil industry is intrinsically linked to the major technological advances that have defined modernity. From the Industrial Revolution in the 18th century to the digital age of the 21st century, the exploitation of energy resources has driven unprecedented economic development. This chapter examines how each historical stage has influenced the industry's evolution, paving the way for the current integration of artificial intelligence (AI).

1. The Industrial Revolution: The First Steps in Energy Use

The Beginning of Fossil Fuel Usage
The Industrial Revolution marked a turning point in global economic history. During this period, coal became the primary fuel driving industrial and transportation mechanization. However, by the late 19th century, the discovery of oil began to reshape the energy landscape.

The First Modern Oil Drilling
In 1859, Edwin Drake drilled the first oil well in Pennsylvania, USA. This event laid the foundation for the modern oil industry, demonstrating that oil could be efficiently extracted to meet growing energy demands.

Global Expansion
As oil became a strategic resource, its exploitation rapidly expanded worldwide, with oil fields discovered in Russia, the Middle East, and Latin America.

2. The 20th Century: The Era of Mass Production

World War I and the Strategic Importance of Oil
During World War I, oil emerged as an essential resource for the war effort, fueling warships, vehicles, and aircraft. This period underscored the need for a steady and reliable oil supply.

The Rise of Major Oil Companies
In the 1920s, the first major oil corporations, known as the "Seven Sisters," emerged. These

companies controlled the exploration, production, and distribution of oil, establishing the framework for the 20th-century global economy.

World War II and the Post-War Oil Boom
Oil demand surged again during World War II. In the post-war period, global economic expansion drove an unprecedented increase in oil consumption, particularly in the transportation sector.

3. The Second Half of the 20th Century: Technological Advances in Exploration and Production

Introduction of Offshore Drilling
In the 1950s, offshore drilling technology enabled oil extraction from marine platforms, opening new frontiers for the industry.

Process Automation
In the 1970s and 1980s, the industry began adopting automation technologies and distributed control systems (DCS) to optimize production and reduce costs.

Oil Crisis and Diversification
The 1973 oil crisis highlighted the vulnerability of economies dependent on crude oil, spurring research into energy efficiency and alternative technologies.

4. The 21st Century: The Digital Era

Advances in Exploration and Production
With the arrival of the new millennium, technologies like horizontal drilling and hydraulic fracturing revolutionized the industry, enabling the exploitation of unconventional reserves such as shale oil.

Digitalization and Big Data
The development of advanced sensors and SCADA systems allowed the massive collection of operational data. These technologies laid the groundwork for the application of AI by providing the necessary data for predictive models and real-time optimization.

5. The Artificial Intelligence Revolution

AI as a Paradigm Shift
Today, AI is not just a tool; it represents a paradigm shift. For the first time, the industry can anticipate failures, optimize operations, and adjust production in real time, with a level of precision and speed beyond human capability.

Comparison with Previous Revolutions
Like the mechanization of the Industrial Revolution and the automation of the 20th century, AI is transforming not only how oil is produced but also how every decision in the process is made, from exploration to distribution.

A Journey of Innovation and Progress
The evolution of the oil industry reflects the story of human innovation. Each stage, from the Industrial Revolution to the digital era, has been a step toward greater efficiency and sustainability. The integration of AI represents the next great leap, solidifying the oil industry's place in a world demanding smarter and more responsible energy solutions. In this context, understanding the past is essential to appreciating AI's transformative impact in the present and future.

Chapter: Future Innovations - The Convergence of Emerging Technologies

Digital transformation in the oil industry goes beyond the implementation of artificial intelligence (AI). Other emerging technologies, such as blockchain and the Internet of Things (IoT), are beginning to play crucial roles, complementing and enhancing AI's capabilities. The integration of these technologies will not only improve operational efficiency but also revolutionize data management, security, and transparency across the value chain.

1. Blockchain: Transparency and Security in the Value Chain

Blockchain, known for its ability to provide an immutable and decentralized record of transactions, has the potential to transform critical aspects of the oil industry.

Applications in the Oil Industry

- **Smart Contract Management**
  Companies can automate agreements between parties, such as supply and transportation contracts, reducing intermediaries and minimizing the risk of contractual disputes.
- **Supply Chain Transparency**
  Blockchain enables real-time tracking of each barrel of oil from extraction to its final destination, ensuring product authenticity and traceability.
- **Security and Fraud Prevention**
  The immutable nature of blockchain protects critical operational data from tampering, ensuring operational and financial information is always secure and verified.

Potential Benefits

- **Reduced Administrative Costs:** Automating traditionally human-dependent processes.
- **Increased Trust and Efficiency:** Enhancing business relationships through transparency.

2. Internet of Things (IoT): Connectivity and Real-Time Monitoring

IoT, through its network of connected devices, enables the real-time collection of data from equipment, infrastructure, and environments.

Applications in the Oil Industry

- **Equipment Monitoring and Predictive Maintenance**
  IoT sensors installed on equipment such as pumps and separators collect real-time performance data. These data are processed by AI systems to identify signs of wear or imminent failure, allowing maintenance before critical failures occur.
- **Optimizing Operations in Remote Wells**
  In regions with difficult access, IoT devices enable remote monitoring and control, reducing the need for physical visits and enhancing worker safety.
- **Environmental Management**
  IoT environmental sensors can measure gas emissions, water quality, and other critical parameters, helping companies comply with environmental regulations and minimize their ecological footprint.

Potential Benefits

- **Greater Operational Efficiency:** Reducing unplanned downtime and optimizing resources.
- **Improved Safety and Regulatory Compliance:** Constant monitoring of hazardous conditions and adherence to regulatory standards.

3. The Convergence of AI, Blockchain, and IoT: An Integrated Ecosystem

The true revolution in the oil industry will occur when these technologies work together as part of an integrated ecosystem.

Integrated Use Case

- **Smart Production and Logistics**
  IoT sensors collect real-time equipment data, which AI processes to optimize production. Simultaneously, blockchain records all transactions and operations to ensure transparency

and security. This system guarantees that crude is produced, transported, and sold with maximum efficiency and without disruptions.

- **Emergency Response**
A spill or critical failure can be instantly detected by IoT sensors. AI analyzes the situation and activates emergency protocols, while blockchain records every action taken, ensuring traceability of the response.

4. Challenges and Requirements for Implementation

Although these technologies offer immense potential, their widespread adoption requires overcoming certain challenges:

- **Interoperability:** Ensuring that AI, IoT, and blockchain systems integrate seamlessly with existing infrastructures.
- **Cybersecurity:** As connectivity expands, so does the risk of cyberattacks. Companies must implement robust protocols to protect their systems.
- **Training and Organizational Culture:** Preparing personnel to operate in a highly digitized environment is key to a successful transition.

A Connected and Transparent Future

The convergence of emerging technologies such as AI, blockchain, and IoT promises to elevate the oil industry to new levels of efficiency, security, and sustainability. These technologies not only optimize existing operations but also open the door to new business models based on transparency and collaboration. Companies that adopt this integrated approach will lead the next era of the energy industry.

Chapter: Geopolitical Implications - Effects of Uneven AI Adoption in the Global Market

The oil industry has historically been a central axis in geopolitical relations. Oil-producing countries, particularly those with vast reserves, have wielded significant power on the global stage. However, the integration of artificial intelligence (AI) into oil operations is generating a paradigm shift that could redistribute this power. Uneven AI adoption has the potential to drastically alter the balance in the global market, favoring those who adapt more quickly.

1. AI as a Geopolitical Differentiator

New Dynamics of Competitiveness
Countries that adopt AI in their oil operations will be better positioned to:

- **Reduce operational costs:** Through predictive maintenance and production optimization.
- **Increase efficiency and production:** Maintaining high extraction levels even in marginal wells.
- **Adapt to global demand:** Quickly adjusting production in response to market fluctuations.

These benefits will allow them to compete at lower prices and capitalize on periods of high demand with superior flexibility.

Case Study: Middle East vs. North America

- **Middle East:** Many countries in this region possess the largest oil reserves, but their technological infrastructure has yet to fully adopt AI. Dependence on traditional methods could limit their long-term competitiveness.
- **North America:** The United States and Canada are at the forefront of AI implementation in oil fields, particularly in shale oil extraction. This gives them a significant competitive edge, enabling profitable production even when crude prices are low.

2. The Technological Divide Between Developed and Developing Countries

AI adoption could deepen the gap between countries with access to advanced technology and those lacking the resources to implement it.

Developed Countries
With access to capital, talent, and technology, these countries will lead AI adoption, consolidating their dominance in the global market. They will be able to:

- **Maximize revenues:** Through more efficient operations.
- **Diversify economies:** By reinvesting profits into technology and sustainability sectors.

Developing Countries
Lacking the resources to invest in AI, these countries will face challenges such as:

- **Decreased competitiveness:** Higher operational costs will exclude them from key markets.
- **Growing dependency:** On developed countries for technology and operational support.
- **Loss of geopolitical relevance:** As their traditional markets shrink.

3. Shifts in Trade Relations and Power

Uneven AI adoption could reshape alliances and rivalries in the energy industry.

Formation of New Geopolitical Blocs

- **Technological Alliances:** Countries leading AI adoption may form strategic alliances, sharing technologies and optimizing joint operations.
- **Export Dependence on Technology:** Nations developing AI solutions specific to the oil industry could export them, establishing new trade and power relations.

Challenges for OPEC and Other International Organizations
The Organization of Petroleum Exporting Countries (OPEC) may struggle to maintain its

influence if its members do not adopt AI uniformly. Technological inequality could create internal tensions, complicating coordination on quotas and pricing.

4. Implications for Energy Security

New Cybersecurity Threats
Reliance on automated and AI-based systems increases vulnerability to cyberattacks. Countries that do not invest in cybersecurity will face greater risks of sabotage in their energy operations.

Diversification of Energy Sources
AI could accelerate the transition to renewable energy sources, reducing global dependence on oil. This shift could diminish the geopolitical relevance of countries traditionally reliant on crude oil exports.

5. Future Projections

If the current trend of technological adoption continues, the next 10 to 20 years could see the geopolitics of oil dominated by:

- **Technologically advanced countries**, controlling most of the market through efficient and sustainable operations.
- **New technological alliances**, competing for leadership in a diversified energy market.

On the other hand, countries that do not adopt AI could face a gradual loss of relevance, with severe implications for their economies and global influence.

Conclusion: AI as a Global Power Reconfigurator

Artificial intelligence is not just an operational tool; it is a paradigm shift redefining the geopolitical rules of the oil industry. Countries leading its adoption will consolidate their power, while those lagging will face an uncertain future. In this context, investing in AI is not an option but a strategic necessity to ensure relevance and success in a dynamic global market.

Tank and Separator Management with AI: Efficiency, Safety, and Sustainability

The management of tanks and separators is crucial in oil fields, enabling the proper separation of oil, gas, and water to ensure product quality and minimize environmental impact. The integration of artificial intelligence (AI) in this area is revolutionizing operations by improving efficiency, safety, and sustainability. Below, the key aspects of this technological transformation are detailed.

1. Level and Composition Control

Precise control of levels and composition in tanks and separators is critical to prevent overflows, maintain operational stability, and ensure product quality.

a. Continuous Level Monitoring

- **High-Resolution Sensors:**
  Tanks and separators are equipped with ultrasonic, capacitive, and differential pressure sensors, which measure liquid and gas levels in real time with millimeter precision.
- **Real-Time Analysis:**
  Data from these sensors are processed by AI algorithms that automatically adjust valves and pumps to maintain levels within safe and optimal ranges.

Technical Example:
If a separator experiences a rapid increase in water level, AI adjusts the water discharge valve and redistributes the flow to other storage tanks, preventing overflow and maintaining stable operations.

b. Automatic Composition Regulation

- **Specific Composition Sensors:**
  Equipment such as inline chromatographs and optical sensors detect the exact proportions of oil, water, and gas in the fluid.

- **Data-Driven Optimization:**
  AI automatically adjusts pressure, temperature, and flow rate in separators to maximize separation efficiency.

Technical Benefit:
This ensures that crude oil has a uniform composition, reducing impurities and improving its commercial value.

c. Automatic Response to Abnormal Conditions

- **Rapid Variation Detection:**
  AI detects fluctuations in the proportion of water or gas in crude oil and immediately adjusts the process to avoid interruptions or quality degradation.

Advantage:
This capability significantly reduces reaction times to unforeseen changes, improving operational reliability.

2. Optimization of the Separation Process

The efficiency of separators is essential for maximizing production and minimizing costs. AI enables continuous adjustment of operational parameters, ensuring optimal separation.

a. Predictive Models for Continuous Optimization

- **Machine Learning Algorithms:**
  Predictive models, trained on historical and operational data, identify ideal conditions to maximize separation efficiency.

Technical Example:
If data suggest that lowering operating pressure would improve gas separation efficiency, AI adjusts the pressure control valve in real time.

b. Energy Consumption Reduction

- **Optimization of Auxiliary Equipment:**
  AI automatically adjusts the operation of heaters and compressors, ensuring they run only when necessary and at optimal levels.

Quantifiable Impact:

This can reduce the energy consumption of separators by up to 20%, lowering operational costs and carbon emissions.

c. Improved Final Product Quality

- **Precise Separation:**
  By operating within optimal parameters, AI minimizes the presence of contaminants such as water and solids in crude oil.

Result:

Cleaner crude reduces wear on pipelines and transportation equipment, lowering maintenance costs and extending asset lifespan.

3. Leak and Spill Prevention

Safety and environmental protection are top priorities in tank and separator management. AI plays a crucial role in detecting and preventing leaks and spills.

a. Early Detection of Structural Anomalies

- **Ultrasonic and Pressure Sensors:**
  Continuously monitor the integrity of structures, detecting signs of corrosion, cracks, or mechanical stress.
- **Predictive Analysis:**
  AI identifies patterns indicating potential structural failure before it occurs.

Technical Example:

A slight pressure drop in a tank may indicate an incipient leak. AI alerts the maintenance team, enabling immediate intervention.

b. Alert Systems and Automatic Response

- **Automatic Containment Activation:**
  Upon detecting an imminent risk, AI shuts valves, stops pumps, and activates containment systems to minimize impact.

Technical Advantage:
This ensures near-instantaneous response, even in remote areas, significantly reducing environmental risks.

c. Environmental Risk Prediction

- **Predictive Risk Models:**
  Based on historical and operational data, AI can predict when and where leaks or spills are most likely to occur.

Benefit:
This capability allows for preventive maintenance planning, reducing the risk of incidents.

Global Impact on Tank and Separator Management

a. Operational Efficiency

- **Continuous Optimization:**
  AI ensures that tanks and separators always operate under optimal conditions, reducing downtime and improving overall productivity.

b. Cost Reduction

- **Energy Savings:**
  AI optimizes the use of heaters and compressors, reducing energy consumption by up to 20%.

- **Proactive Maintenance:**
  By preventing failures and leaks, emergency repair costs and environmental fines are minimized.

c. Regulatory Compliance and Sustainability

- **Carbon Footprint Reduction:**
  Energy optimization and leak prevention contribute to compliance with strict environmental regulations, improving operational sustainability.

d. Operational Safety

- **Protection of Workers and Assets:**
  By minimizing exposure to hazardous environments and preventing major incidents, AI significantly enhances personnel safety and asset integrity.

Conclusion: The Transformative Role of AI in Tank and Separator Management

The integration of artificial intelligence (AI) in tank and separator management marks a revolutionary shift in the oil industry. This technological advancement not only optimizes key processes but also redefines operational, safety, and sustainability standards.

1. Operational Efficiency: A New Level of Precision

AI enables continuous and optimized operation through real-time monitoring, automatic regulation, and advanced predictive modeling.

- **Proactive Optimization:**
  AI algorithms dynamically adjust parameters such as pressure, temperature, and flow rate to keep separators and tanks at their optimal operating point, reducing resource consumption and maximizing efficiency.
- **Intelligent Monitoring:**
  Advanced sensors and predictive analysis work together to detect and correct deviations

before they escalate, ensuring a stable and efficient separation process even under fluctuating conditions.

2. Cost Reduction and Resource Maximization

AI's ability to predict and prevent failures, adjust energy consumption, and optimize auxiliary equipment operations directly impacts operational costs.

- **Significant Energy Savings:**
  Dynamic optimization can reduce energy consumption by up to 20%.
- **Proactive Maintenance and Reduced Downtime:**
  AI identifies signs of wear and potential failures weeks in advance, allowing maintenance to be scheduled during low-demand periods, minimizing unplanned interruptions and extending equipment lifespan.

3. Operational Safety and Environmental Protection

AI enhances operational safety by detecting early structural anomalies and risks of leaks, enabling rapid and effective intervention.

- **Prevention of Major Incidents:**
  Continuous monitoring ensures prompt responses to potential risks, mitigating the impact of spills or explosions.
- **Environmental Compliance:**
  AI optimizes energy use, reduces emissions, and prevents leaks, ensuring compliance with strict environmental regulations and improving public perception.

4. Sustainability and Future Adaptation

AI implementation in tanks and separators significantly contributes to operational sustainability, aligning operations with global sustainability goals.

- **Carbon Footprint Reduction:**
  Reduced energy consumption and spill prevention directly contribute to lower emissions.

- **Future-Ready Operations:**
  AI's learning and continuous improvement capabilities ensure operations adapt to changes in demand, new regulations, and technological advancements, positioning companies as leaders in innovation and environmental responsibility.

5. Competitive Advantage and Market Resilience

In an environment where competition and sustainability expectations are increasing, AI offers companies a crucial strategic advantage.

- **Operational Flexibility:**
  The ability to quickly adjust operations based on well conditions and market demands allows companies to maximize revenue and minimize risk.
- **Reputation and Trust:**
  Companies adopting AI not only operate more efficiently and sustainably but also enhance their public image, earning the trust of regulators, investors, and communities.

The implementation of AI in tank and separator management is a milestone in the evolution of the oil industry. By integrating advanced monitoring, predictive analysis, and optimization technologies, AI ensures safer, more efficient, and sustainable operations. This approach not only boosts profitability by reducing costs and maximizing production but also guarantees compliance with environmental regulations and reinforces operational safety.

Companies adopting this technology will be better prepared to face future challenges, from market volatility to growing pressure for cleaner and more responsible operations. AI is not just an operational tool but a catalyst for a more resilient and sustainable industry capable of leading in an increasingly demanding global market.

Real-Time Controls and Automation with AI: Transforming Supervision and Operation in Oil Fields

The integration of artificial intelligence (AI) into process supervision and control through Distributed Control Systems (DCS) and Supervisory Control and Data Acquisition (SCADA) systems is revolutionizing the oil industry. These systems, essential for automated operations, have been enhanced by AI to offer smarter, more adaptive, and efficient control. Below, we break down the specific capabilities AI brings to these environments, highlighting its technical and operational impact.

1. Integrated Supervision Systems (DCS, SCADA) with AI

a. Automation of Complex Processes

AI transforms traditional control systems by enabling autonomous and dynamic management of critical processes, reducing the need for human intervention.

- **Intelligent Management of Operational Variables:**
  AI simultaneously analyzes hundreds of operational variables such as pressure, temperature, flow, and fluid composition. This allows for the automatic adjustment of valves, pumps, and other devices to maintain optimal conditions.

Technical Case:
In a three-phase separator, if gas pressure exceeds a critical threshold, AI immediately adjusts the release valve and regulates inflow rates to prevent overflows or efficiency losses.

b. Predictive and Preventive Analysis

- **Early Anomaly Detection:**
  AI, trained with historical and real-time data, identifies patterns that precede failures or inefficiencies, providing early warnings of potential issues.

Technical Example:

A predictive AI model detects a rising trend in compressor temperature, indicating progressive wear. AI suggests maintenance before a catastrophic failure occurs, preventing costly downtime.

c. Intelligent Redundancy and Operational Continuity

- **Automated Management of Alternative Control Paths:**
  AI ensures that if one operational route fails, the system can automatically redirect operations to redundant paths without interruption.

Technical Impact:

This is crucial in critical operations where even brief interruptions could lead to significant losses or environmental damage.

2. Automatic Adjustments to Changing Conditions

AI's ability to respond to real-time variations is a key differentiator from traditional systems.

a. Dynamic Response to Operational Changes

- **Adaptation in Wells with Variable Behavior:**
  AI continuously monitors well fluctuations, such as reservoir pressure changes, free gas production, or crude viscosity variations, automatically adjusting flow rates and separator parameters.

Practical Case:

If a well begins producing more free gas, AI reduces separator pressure to prevent foaming, improving separation efficiency and preventing blockages.

- **Compensation for Extreme Environmental Conditions:**
  Climatic factors such as sandstorms, heatwaves, or freezes can negatively impact equipment. AI automatically adjusts pump speeds, valve openings, and compressor loads to protect equipment and maintain safe operations.

Technical Example:

During a heatwave, AI detects an increase in a pump motor's temperature and reduces its speed to prevent overheating, ensuring equipment integrity.

b. Response Based on Market Demand

- **Optimization According to Prices and Demand:**
  By integrating real-time market data, AI adjusts production rates to maximize profitability.

Use Case:

If crude oil prices rise, AI increases production within safe limits. Conversely, during periods of oversupply or low prices, it reduces operations to avoid overloading storage tanks.

- **Dynamic Inventory Management:**
  AI monitors storage levels and adjusts extraction and production rates to maintain optimal inventory levels.

3. Key Benefits of Real-Time Automation

a. Safer Operations

AI significantly reduces operational risks by responding faster than human operators in critical situations.

- **Emergency Prevention:**
  By detecting anomalies such as leaks, overpressure, or overheating, AI automatically activates containment and mitigation systems, protecting both personnel and the environment.

b. Energy Efficiency

AI continuously optimizes energy use, ensuring equipment operates at peak efficiency.

- **Energy Consumption Optimization:**
  AI dynamically adjusts power levels in pumps, compressors, and heaters to minimize consumption without compromising operations.

Quantifiable Impact:
This can reduce energy consumption by 15-25%, contributing to lower operational costs and a reduced carbon footprint.

c. Operational Continuity and Resilience
AI ensures operational continuity even under challenging conditions, such as extreme demand fluctuations or equipment failures.

- **Intelligent Operational Redundancy:**
  By automatically managing alternative control paths, AI minimizes downtime and maximizes system availability.

d. Workforce Optimization
With advanced automation, human personnel can focus on strategic, high-value tasks such as operational planning and continuous improvement.

- **Reduced Operational Workload:**
  AI takes over routine and repetitive tasks, allowing operators to become analysts and strategists, improving decision-making and innovation.

Long-Term Industry Impact

a. Transformation of Human Roles
AI does not eliminate the need for personnel but redefines operational roles.

- **Data and AI Engineers:**
  Essential for developing, implementing, and maintaining AI systems, ensuring their continuous adaptation to operational needs.

- **Remote Operations Supervisors:**
  While AI automates many tasks, human supervision will remain key in strategic decision-making and managing unforeseen exceptions.

b. Global Competitiveness and Operational Resilience

Companies that adopt AI in their control systems will gain a significant competitive advantage.

- **Superior Operational Efficiency:**
  The ability to respond quickly to operational and market changes positions these companies as leaders in efficiency.
- **Adaptability to Regulations and Sustainability:**
  With the capability to continuously optimize resources and reduce environmental impact, companies will be better prepared to meet increasingly stringent regulations.

Conclusion: Enhanced Real-Time Control and Automation with AI

The integration of artificial intelligence with DCS (Distributed Control Systems) and SCADA (Supervisory Control and Data Acquisition) is transforming the oil industry, setting a new standard in operational management. This fusion of technologies enables an unprecedented level of control and automation, driving efficiency, safety, and sustainability in oil fields.

AI not only enhances supervisory and control capabilities but also introduces critical features such as failure prediction, dynamic adaptation to operational and market conditions, and real-time energy optimization. These capabilities ensure that operations remain within optimal limits, even in highly volatile and challenging environments.

1. Transformational Approach to Safety and Efficiency

AI redefines operational safety by detecting and responding to critical situations faster than any human intervention. From early leak detection to preventing overpressure and mechanical failures, AI acts as an intelligent safety barrier, protecting workers and the environment. Simultaneously, energy consumption optimization and continuous process improvement enable more efficient operations, with significant cost savings.

2. Operational Continuity and Resilience

AI's ability to automatically manage operational redundancies ensures continuity even under adverse conditions, minimizing downtime and maximizing system availability. This is crucial in an industry where every minute of downtime can translate into millions in losses.

3. Redefining Human Roles and Global Competitiveness

Rather than replacing human labor, AI redefines roles in the industry, focusing on strategic, high-value tasks. Data and AI engineers, along with remote operations supervisors, will be crucial in ensuring the efficiency and adaptability of these advanced systems.

Additionally, companies adopting this technology will be better positioned to compete in a global market characterized by volatility and strict environmental regulations. The combination of operational efficiency, sustainability, and responsiveness ensures not only short-term viability but also sustained leadership in the industry.

4. Preparing for the Future

Implementing AI in real-time automation is more than a technological advancement; it is a strategic investment for the future. As the industry faces challenges such as price fluctuations, energy transition, and regulatory pressure, AI becomes an essential pillar for ensuring resilient, sustainable, and highly competitive operations.

Artificial intelligence is redefining supervision and control in oil fields, delivering a level of precision, adaptability, and efficiency that was unattainable with traditional systems. By enhancing safety, optimizing performance, and ensuring operational continuity, AI positions itself as an indispensable tool for the industry. Companies that adopt this technology will not only secure their competitiveness in the present but will also lead the transition toward a more sustainable and technologically advanced future in the energy sector.

Can AI Revolutionize Exploration and Drilling in the Future?

The integration of artificial intelligence (AI) in the oil industry has already proven its effectiveness in areas such as production and predictive maintenance. However, a pivotal question arises: Can AI play a central role in the exploration and drilling of oil reservoirs in the future? This inquiry not only highlights a potential technological leap but also unveils a spectrum of possibilities on how technology can transform some of the industry's most complex and costly processes.

AI-Driven Geological Exploration: A Near Future

Hydrocarbon exploration involves analyzing vast volumes of geological, geophysical, and seismic data to identify potential reservoirs. AI is already demonstrating its potential in this area by improving accuracy and reducing the time required to pinpoint prospective locations.

1. Advanced Analysis of Geological and Seismic Data

- **Enhanced Seismic Imaging Processing:**
  AI algorithms can analyze large seismic data sets in real time, detecting patterns that traditional methods might overlook. This allows for more precise identification of geological formations.
- **Predictive Modeling:**
  Using historical data, AI can predict the location and quality of reservoirs, reducing uncertainty and risk in exploration.

2. Cost and Time Reduction

- **Resource Optimization:**
  By quickly identifying high-potential areas, AI reduces the need for multiple exploratory studies, saving time and resources.
- **3D Simulations and Modeling:**
  AI can generate three-dimensional simulations of underground formations, enabling geologists to virtually assess a reservoir's viability before drilling.

Autonomous Drilling with AI: A Viable Horizon

Drilling is among the most expensive and high-risk operations in the oil industry. Here, AI could have a transformative impact by automating complex processes and enhancing safety and efficiency.

1. Real-Time Drilling Optimization

- **Automated Drilling Parameter Control:**
  AI can automatically adjust drilling speed, mud pressure, and drill bit direction based on real-time data, optimizing the process and minimizing risks like wellbore collapse or lost circulation.
- **Predictive Drilling:**
  Leveraging historical and real-time data, AI can anticipate issues such as high-pressure zones or lithology changes, enabling immediate and precise responses.

2. Enhanced Safety and Risk Reduction

- **Continuous Well Monitoring:**
  Advanced sensors integrated with AI systems provide constant monitoring of well conditions, detecting and responding to potential failures or risks, such as blowouts, before they occur.
- **Autonomous Safety Systems:**
  AI could activate safety mechanisms, such as blowout preventers (BOPs), during emergencies, reducing reaction times and protecting both workers and the environment.

3. Remote Drilling and Operational Autonomy

- **Autonomous Drilling Platforms:**
  In the future, AI could operate drilling platforms entirely autonomously, reducing the need for personnel in remote or hazardous locations.

- **Cost and Logistics Optimization:**
  By minimizing human intervention and automating processes, AI could significantly cut drilling costs and improve operational efficiency.

Challenges and Ethical Considerations

While the implementation of AI in exploration and drilling offers numerous advantages, it also raises significant challenges:

- **Technological Adoption and Training:**
  Transitioning to AI-driven systems requires substantial investment in technology and the continuous training of specialized personnel.
- **Cybersecurity Risks:**
  Automation increases exposure to cyber threats, making the development of robust security systems imperative.
- **Responsibility and Ethics:**
  The autonomous decision-making capability of AI systems in critical situations raises questions about accountability in case of failures or accidents.

AI as the Future of Oil Exploration and Drilling

The question "Can AI revolutionize exploration and drilling in the future?" not only highlights a potential technological breakthrough but also emphasizes a critical transformation in how the oil industry operates its most complex and pivotal processes. The answer points to a resounding yes, based on the significant impact AI is already having in areas like production and predictive maintenance. However, its application in exploration and drilling promises to be even more transformative.

1. Towards More Accurate and Efficient Geological Exploration

Hydrocarbon exploration is inherently risky, costly, and data-dependent. With AI, this paradigm is changing:

- **Deep Analysis of Geological and Seismic Data:**
  AI processes massive volumes of geophysical and seismic data with unmatched speed and precision, identifying patterns and reservoir features that traditional methods might miss.
- **Reduced Exploration Time:**
  Automation and advanced analysis enable faster decision-making, significantly cutting down the time required to identify and assess prospects. This can reduce projects that traditionally take years to mere weeks or months.
- **Minimizing Risks and Optimizing Resources:**
  By focusing operations on areas with higher success probabilities, AI lowers financial and operational risks, optimizing the use of human and technical resources.

2. Autonomous and Safer Drilling: A Future to Realize

Drilling is one of the most complex and risky processes in the oil industry. With AI, the outlook becomes much more promising:

- **Real-Time Optimization:**
  AI-powered autonomous drilling systems continuously adjust operational parameters like drilling speed, mud pressure, and drill bit inclination based on real-time data. This not only improves efficiency but also reduces equipment wear and failure risks.
- **Human Risk Reduction:**
  In operations where conditions are extreme and hazardous, such as deep-water or Arctic drilling, AI could fully control drilling platforms, minimizing personnel exposure to risks.
- **Prevention of Catastrophic Failures:**
  AI's predictive capabilities can anticipate problems like blowouts or lost circulation, automatically activating safety mechanisms to protect both assets and the environment.

3. Global Impact: A Paradigm Shift in the Industry

The integration of AI in exploration and drilling not only results in operational benefits but also impacts strategic and environmental levels:

- **Enhanced Sustainability:**
Precise and optimized drilling reduces the number of exploratory wells needed, minimizing environmental disturbance and lowering the carbon footprint associated with these operations.
- **Operational Cost Reduction:**
Automating tasks and reducing downtime translate into significant savings. In drilling, each additional day can cost millions of dollars, so any efficiency improvement directly impacts profitability.
- **Greater Operational Resilience:**
AI's ability to operate in challenging environments and make autonomous decisions ensures operational continuity even in adverse conditions, strengthening companies' resilience to unexpected events.

4. Overcoming Challenges

Despite its promises, adopting AI in exploration and drilling faces significant obstacles that must be addressed:

- **Investment in Infrastructure and Training:**
Transitioning to autonomous drilling requires substantial investment in technological infrastructure and the training of skilled personnel to operate and oversee these advanced systems.
- **Cybersecurity:**
Dependence on autonomous and connected systems increases vulnerability to cyberattacks. Implementing robust security protocols will be essential to protect critical operations.
- **Acceptance and Regulation:**
The adoption of autonomous technologies in critical and highly regulated sectors like exploration and drilling may face resistance from regulators and communities. Demonstrating these technologies' safety and environmental benefits will be crucial.

5. Competitive Advantage and the Industry's Future

Companies that lead the implementation of AI in exploration and drilling will be strategically positioned to dominate the market. The ability to make faster, more accurate, and data-driven decisions will allow them to reduce costs and improve operational efficiency and sustainability.

- **Leadership in Innovation:**
  Early adopters of AI for exploration and drilling will set the standard, positioning themselves as pioneers in a more technological and sustainable industry.
- **Adaptation to Energy Transition:**
  In a world moving towards greater sustainability, efficient and low-impact drilling will be a key differentiator.

AI has the potential to revolutionize exploration and drilling in the oil industry, making these processes safer, more efficient, and more sustainable. While technical, operational, and regulatory challenges persist, the advantages are undeniable. From reducing risks and costs to improving precision and sustainability, AI is poised to become the cornerstone of the industry's next evolution.

Answering the question, "Can AI revolutionize exploration and drilling in the future?" with a resounding yes is not merely an act of faith in technology but a projection based on the tangible impact already beginning to materialize. Companies that embrace this approach will not only secure their relevance in a competitive market but also lead a transformation that will define the future of the global energy sector.

Increasing Operational Efficiency Through AI: A Detailed Quantitative and Technical Analysis

Artificial intelligence (AI) has become a key catalyst for optimizing operational efficiency in the oil industry. Its ability to analyze real-time data, automate processes, and predict failures enables companies to maximize production, reduce operational (OPEX) and capital (CAPEX) costs, and significantly enhance profitability. Below is a breakdown of the specific benefits of AI integration, supported by quantitative and technical examples.

1. Reducing Production Losses

AI minimizes unplanned downtime, ensuring operations remain at optimal levels.

Quantifiable Impact

- **Reduction in Unplanned Downtime:**
  Implementing AI for predictive monitoring and proactive maintenance can reduce unplanned downtime by 30-50%. In an industry where every hour of downtime represents significant losses, this is critical.

Example:
For an oil field producing 10,000 barrels per day (bpd), a 30% reduction in downtime could result in an additional 1,095,000 barrels annually.

- **Direct Financial Impact:**
  At an average oil price of $75 per barrel, this production increase translates to $82.1 million in additional annual revenue.

Enhanced Operational Efficiency

- **Optimizing Equipment Performance:**
  AI ensures equipment consistently operates near its optimal efficiency point, increasing production by 5-10% by eliminating losses from suboptimal operation.

2. Cost Savings (OPEX and CAPEX)

AI optimizes both operational and capital expenses, improving the financial sustainability of operations.

a. Reducing OPEX (Operational Costs)

- **Automation of Routine Processes:**
  By automating tasks such as equipment monitoring, valve management, and operational parameter adjustments, AI reduces on-site personnel requirements, potentially lowering labor costs by 20-30%.
- **Predictive Maintenance and Failure Reduction:**
  AI anticipates failures and schedules interventions before they occur, reducing maintenance costs by 10-20% and extending equipment lifespan.

Quantifiable Example:
For a field with an annual OPEX of $50 million, these savings could amount to $5-10 million per year.

b. Optimizing CAPEX (Capital Costs)

- **Extending Equipment Lifespan:**
  AI-driven efficient operations reduce equipment wear and tear, extending the lifespan of assets like ESP pumps by 15-20%.

Quantifiable Example:
If the average replacement cost of an ESP pump is $150,000, extending the lifespan of 100 pumps by 20% could save $3 million in CAPEX over an operational cycle.

- **Data-Driven Investment Prioritization:**
  AI helps companies make informed decisions on where and when to invest, avoiding unnecessary expenditures and maximizing returns on capital investments.

3. Time Savings and Improved Decision-Making

AI transforms operational decision-making, reducing response times and improving decision quality.

Real-Time Decisions

- **Faster Response Times:**
  Decisions that previously took hours or days can now be executed within seconds, thanks to AI algorithms.

Technical Example:
In cases of pressure fluctuations indicating imminent blockages, AI can automatically adjust flow rates and notify operators, preventing potential damage.

Real-Time Process Optimization

- **Continuous and Automated Adjustments:**
  AI monitors and adjusts operational parameters such as pressure, temperature, and flow without human intervention, saving 50-70% of time on critical processes.

Total Operational Efficiency Impact:
Time savings and optimized decisions can boost overall field efficiency by 10-15%.

Quantifiable Example:
For a field with an annual EBITDA of $100 million, this efficiency increase could generate an additional $10-15 million in profits.

4. Global Impact on Profitability

The combination of increased production, reduced operational costs, and improved decision-making significantly enhances operational profitability.

Increase in Operational Profitability (ROO)

- **Combined OPEX Reductions:**
  Operational cost savings can range from 15-25%, depending on the level of automation and AI integration.
- **Increase in Effective Production:**
  Maintaining optimal operations can boost production by an additional 5-10%.
- **Overall Profitability Improvement:**
  These combined factors can increase operational profitability (ROO) by 15-20%, strengthening the company's financial position.

Improved Operating Profit Margins

- **Enhanced Competitiveness:**
  By reducing costs and improving efficiency, companies can enhance their operating margins by 10-15%, making them more resilient in volatile pricing environments.

Return on Investment (ROI) in AI Technology

- **Investment Recovery Period:**
  Initial investment in AI systems can be recovered within 2-3 years, with an average annual return of 30% through operational savings and increased revenue.

Financial Example:

An investment of $10 million in AI could yield $3 million in annual benefits after the recovery period.

AI as a Pillar in Increasing Operational Efficiency

The integration of artificial intelligence (AI) in the oil industry is redefining operational efficiency standards, providing tangible benefits across all process stages. From optimizing equipment to enhancing decision-making, AI not only maintains production at optimal levels but also transforms how operations are planned and executed.

1. Profound Impact on Production and Cost Reduction

AI's ability to reduce downtime by 30-50% ensures operational continuity and boosts effective production. In high-production fields, this could result in millions of dollars in additional annual revenue. Similarly, significant savings in OPEX and CAPEX—up to 25% combined—directly improve operational profitability (ROO), bolstering financial strength.

2. Real-Time Decision-Making: A Strategic Advantage

AI enables near-instant decision-making based on real-time data, crucial in an industry where every second counts. Automatic, continuous adjustments to operational parameters maximize equipment efficiency and minimize wear, extending asset life. This approach not only optimizes daily operations but also prepares companies to respond swiftly to market changes and challenging operating conditions.

3. Sustainable Profitability and Global Competitiveness

AI implementation offers remarkable ROI, with a recovery period of 2-3 years and an average annual return of 30%. Additionally, improved energy efficiency and reduced operational costs enable companies to enhance their operating margins by 10-15%, ensuring competitiveness even in volatile pricing environments. This positions companies as not only operational leaders but also key players in a global market demanding efficiency and sustainability.

4. Preparing for the Future

AI is not just an operational tool; it's a strategic enabler positioning companies to tackle future industry challenges. With a focus on sustainability, resilience, and adaptability, organizations adopting AI technologies will be better equipped to meet stricter regulations, respond to pressure for reduced carbon footprints, and capitalize on opportunities in an ever-evolving market.

AI has become indispensable for enhancing operational efficiency in the oil industry. Its ability to optimize production, reduce costs, and improve decision-making transforms not only daily operations but also long-term strategy. Companies that embrace these advanced technologies will be positioned to lead the industry, benefiting from more profitable, sustainable, and competitive operations in an increasingly challenging global environment. AI is not just the future of the oil industry; it is the key to its ongoing success and evolution toward a smarter, more resilient model.

Macroeconomic Impact of AI Implementation in the Oil Industry

The widespread adoption of artificial intelligence (AI) in the oil industry will have significant macroeconomic effects, reshaping the dynamics of the global energy market and affecting countries and companies in various ways. Below is an analysis of these impacts and the potential consequences for those that fail to implement these technologies.

1. Redistribution of Global Competitiveness

Countries and Companies Adopting AI

- **Cost Reduction and Greater Efficiency:**
  AI enables companies to reduce operational costs (OPEX) by up to 25% and optimize capital expenditures (CAPEX), resulting in reduced reliance on high oil prices to maintain profitability. This will strengthen the competitiveness of countries and companies adopting AI, even in low-price environments.
- **Strengthening Export Economies:**
  Oil-producing countries implementing AI can maintain positive margins at lower prices, ensuring their position as key suppliers in the global market. This could increase their share of international trade, improve their trade balance, and strengthen their currencies.

Countries and Companies Not Adopting AI

- **Declining Competitiveness:**
  Those failing to adopt AI will face higher production costs, forcing reliance on oil prices above $50-60 per barrel to remain profitable. This will make them less competitive in saturated or low-price markets.
- **Loss of Market Share:**
  Less efficient companies may be outcompeted by those operating with lower margins, directly affecting oil-producing countries' ability to attract foreign investment and sustain stable export revenues.

2. Transformation of the Global Energy Balance

Promotion of Cleaner Energy Sources

AI facilitates the integration of renewable energy into oil operations, reducing reliance on fossil fuels and lowering the carbon footprint. This could accelerate the global energy transition, with the following impacts:

- **Reduced Oil Demand:**
  As renewable energy production costs decrease and integration improves, global oil demand may stabilize or even decline in the long term, particularly affecting countries heavily reliant on oil revenues.
- **Competition in Emerging Energy Industries:**
  Countries investing in AI technologies will not only ensure efficiency in hydrocarbon production but also position themselves to compete in emerging sectors such as carbon capture and storage (CCS) and green hydrogen.

Impact on Oil-Dependent Countries

- **Risk of "Carbon Lock-In":**
  Countries with large oil reserves and economies heavily reliant on these resources but without access to or investment in AI may find themselves stuck with uncompetitive assets. This could accelerate the "stranded assets" phenomenon, where reserves lose value before being exploited due to declining global demand.

3. Economic Inequality Among Oil Producers

Technological and Economic Gap

Uneven AI implementation could widen the gap between technologically advanced countries and companies and those lacking the resources to adopt these innovations.

- **Advanced Economies:**
  Countries with access to capital, infrastructure, and technological talent will optimize

their industries, attract more foreign direct investment (FDI), and consolidate their position in the global market.

- **Developing Countries:**
Oil producers in emerging economies that fail to invest in AI could fall behind, losing market share and seeing reduced export revenues. This may exacerbate global economic inequality and increase dependence on less competitive sectors.

Geopolitical Consequences

- **Redistribution of Energy Influence:**
Countries like the United States, Saudi Arabia, and Norway, already advancing in digitizing their energy industries, will solidify their roles as key players in energy geopolitics. Conversely, nations such as Venezuela or Iraq, which face greater challenges in modernizing their industries, may lose international relevance.

4. Pressure on Labor Markets

Transformation of Employment in the Industry

AI will automate many operational tasks, reducing the need for labor in specific areas but also creating demand for new technological roles.

- **Worker Displacement:**
In companies and countries not adopting AI, traditional operational jobs will be at risk due to declining competitiveness. The lack of technological adaptation could lead to a significant reduction in the energy sector workforce.
- **New Employment Opportunities:**
Companies adopting AI will require highly skilled professionals in areas such as data analysis, AI development, and cybersecurity. This will drive the creation of high-specialization jobs, generating new dynamics in labor markets.

Social and Economic Consequences

- **Resilient Economies:**
  Countries investing in education and technological training will mitigate the social impact of labor displacement, diversify their economies, and strengthen resilience.
- **Social Tensions:**
  In oil-dependent regions that fail to adopt advanced technologies, job losses and declining revenues could lead to social instability and political tensions.

Conclusion: The Transformative Impact of AI on the Global Oil Economy

The adoption of AI in the oil industry will have a profound and multifaceted impact on the global economy. Companies and countries that integrate these technologies will not only gain competitiveness but also lead the transition to a more efficient and sustainable energy model. In contrast, those that fail to adopt AI will face higher costs, loss of market share, and economic pressure, potentially diminishing their relevance in global energy geopolitics.

Key Takeaways:

1. **AI as a Catalyst for Global Competitiveness:**
   Reducing operational costs by up to 25% and maintaining profitability at lower oil prices ensures that AI adopters remain competitive, even in volatile environments.
2. **Leadership in Sustainability:**
   AI optimizes operational efficiency while promoting sustainable practices, positioning companies as leaders in reducing emissions and integrating renewables.
3. **Dynamic Adaptability in Volatile Markets:**
   AI enables companies to predict demand with over 90% accuracy, adjust production in real-time, and optimize resource use, leading to potential annual revenue increases of 5-10%.
4. **Socioeconomic and Geopolitical Implications:**
   Unequal AI adoption will widen the gap between advanced and developing economies, influencing labor markets, investment flows, and geopolitical power dynamics.

Strategic Imperative:

The implementation of AI in the oil industry is no longer an optional strategy but a necessity to ensure long-term survival and success. Companies and countries that embrace this technological revolution will be better prepared to capitalize on opportunities and mitigate risks associated with the global energy transition. Those who fail to adapt risk becoming marginalized in an increasingly digital and sustainable global economy.

Redefining Job Roles in the Oil Industry with AI: A Technical Perspective

The implementation of artificial intelligence (AI) in oil fields is transforming the labor structure, shifting repetitive operational tasks toward more technical and strategic roles. Below is a detailed analysis of this evolution in terms of required skills, operational benefits, and new responsibilities.

1. Essential Human Roles: Technical Redefinition

a. Automation and Control Engineers

Required Skills and Technical Knowledge:

- **Design of Automated Systems:**
  Proficient in advanced technologies such as PLCs (Programmable Logic Controllers) and RTUs (Remote Terminal Units) integrated with DCS and SCADA systems.
  Skilled in industrial programming languages such as Ladder Logic, FBD (Function Block Diagram), and Python for developing specific control algorithms.
- **Integration of AI with Operational Processes:**
  Develop and implement AI models for predictive and adaptive control, including Machine Learning algorithms (supervised and unsupervised learning) for real-time operational adjustments.

Key Responsibilities:

- Configure automated systems to manage flow, pressure, and temperature in real time.
- Validate and fine-tune AI configurations to ensure automated decisions align with operational and safety policies.

Operational Impact:
Proper design and implementation of these systems can reduce human error and improve operational efficiency by 15-20%.

b. Petroleum Data Specialists

Required Skills and Technical Knowledge:

- **Big Data Management and Predictive Modeling:**
  Expertise in platforms like Apache Hadoop, Spark, and NoSQL databases for handling large volumes of operational data.
  Proficiency in advanced analytics tools such as MATLAB, R, and Python (Pandas, NumPy, TensorFlow libraries).
- **Predictive Modeling:**
  Develop and refine predictive models based on historical and real-time data, continually adjusting operational recommendations.

Key Responsibilities:

- Analyze operational trends to identify bottlenecks and propose improvements.
- Develop customized dashboards for visualization and decision-making.

Operational Impact:

Efficient analysis can enhance decision-making accuracy by 90%, reducing operational failures and optimizing production.

c. Advanced Maintenance Technicians

Required Skills and Technical Knowledge:

- **Diagnosis and Repair of Automated Systems:**
  Experience with equipment like ESP and PCP pumps, separators, and automated valves. Knowledge of predictive monitoring tools such as vibration analysis and infrared thermography.
- **Proactive and Predictive Maintenance:**
  Ability to interpret AI-generated data and perform precise interventions before major failures occur.

Key Responsibilities:

- Execute predictive maintenance recommendations from AI systems to ensure operational continuity.
- Recalibrate and adjust sensors and actuators based on operational conditions.

Operational Impact:
Timely intervention can extend equipment lifespan by 15-25%, significantly reducing CAPEX.

d. Field Supervisors with AI Knowledge

Required Skills and Technical Knowledge:

- **Supervision of Automated Operations:**
  Familiarity with SCADA, DCS systems, and remote monitoring software. Understanding AI operational logic and ability to manually intervene during critical conditions.
- **Real-Time Anomaly Resolution:**
  Interpret real-time operational data and make swift decisions when AI detects anomalies.

Key Responsibilities:

- Ensure automated operations remain within optimal parameters.
- Intervene during critical scenarios where human judgment is essential to avoid operational risks.

Operational Impact:
An efficient supervisor can improve response to critical situations by 50-70%, minimizing disruptions and risks.

2. Decline of Routine Roles: Full Automation of Operational Tasks

a. Manual Operators

Tasks Eliminated or Reduced:

- Manual monitoring of indicators on pumps, separators, and tanks.
- Manual adjustments of valves and flow controls.

Role Shift:

These operators can transition to technical assistant roles in remote control centers or advanced monitoring tasks, requiring training in digital tools.

b. On-Site Shift Supervisors

Tasks Eliminated or Reduced:

- Continuous on-site supervision of routine operations.
- Manual review of operational parameters during shifts.

Role Shift:

Supervisors now manage operations from remote control centers, with AI sending critical alerts. This transition enables more strategic and less operational oversight.

Cost Impact:

Automation can reduce on-site personnel needs by 20-30%, leading to significant OPEX savings.

3. Global Impact on Labor Structure

New Role Distribution:

- **AI Technicians and Analysts:** 40% of technical staff.
- **Remote Supervisors and Strategists:** 30%.
- **Reduced Manual Roles:** 30%, focusing on exception monitoring tasks.

4. Continuous Training and Development: Key to a Successful Transition

The shift to an AI-driven oil industry necessitates significant changes in workforce preparation and competencies. Adopting advanced technologies transforms operational processes and redefines job roles, requiring reskilling and continuous development. Companies must prioritize technical training programs to ensure their workforce is equipped to maximize the potential of these technologies.

1. Training in AI and Advanced Automation

Understanding AI systems and automation is essential for modern operations in the oil industry.

- **AI Fundamentals in the Industry:**
  Workers must understand how AI systems process data, detect patterns, and make operational decisions. This includes the functioning of machine learning algorithms and their application in process optimization.
- **Operational Process Automation:**
  Training in configuring and supervising automated systems, such as valve control, pump operation, and separator management through AI. This knowledge enables employees to interpret AI-generated outcomes and act accordingly.

Expected Impact:
Workers trained in these areas will collaborate effectively with AI, ensuring automated operations remain within optimal parameters, reducing failure risks, and improving overall efficiency.

2. Data Analysis and Predictive Modeling

The ability to analyze data and build predictive models will be crucial as decision-making increasingly relies on information generated by sensors and automated systems.

- **Handling Large Data Volumes:**
  Employees must learn to work with real-time data from multiple sources, such as pressure, temperature, and vibration sensors.
- **Predictive Modeling:**
  Training in creating and interpreting predictive models to anticipate failures, optimize production, and enhance maintenance planning.

Technical Benefit:

The ability to interpret these models enables companies to predict problems before they occur, minimizing unplanned downtime and extending equipment lifespan.

3. Maintenance of Automated Equipment

With AI adoption, maintenance shifts from reactive to proactive and predictive approaches.

- **Diagnosis and Troubleshooting in Automated Systems:**
  Technicians must diagnose and resolve issues in highly automated equipment, such as ESP pumps, smart valves, and separation systems.
- **Predictive Maintenance Scheduling:**
  Training in scheduling and executing maintenance tasks based on predictive system recommendations, using operational data to anticipate failures.

Operational Effect:

Optimized predictive maintenance reduces operational costs and enhances safety by preventing catastrophic failures.

The Evolution of the Workforce in the Oil Industry with AI

The implementation of AI not only enhances technical efficiency but also transforms the workforce structure. This shift has significant implications for profitability, sustainability, and global competitiveness.

1. Reallocation of Roles: From Operational to Strategic

Automation of repetitive tasks frees workers from operational roles, allowing them to focus on higher-value activities:

- **Strategic Supervision:**
  Supervisors can now concentrate on operational planning and strategic decision-making, leveraging AI-generated insights to enhance field efficiency and profitability.
- **Data Engineers and Analysts:**
  Analyzing large volumes of data and interpreting predictive models become essential tasks for maximizing AI's value.

2. Driving Innovation and Productivity

Reskilling the workforce not only ensures operational continuity but also fosters innovation:

- **Culture of Innovation:**
  Employees trained in advanced technologies are better equipped to identify and propose data-driven operational improvements.
- **Increased Productivity:**
  With a technologically empowered workforce, companies can implement operational changes more rapidly, boosting productivity and enhancing competitiveness.

3. Reducing Environmental Footprint and Enhancing Sustainability

AI, combined with a skilled workforce, enables companies to minimize their environmental impact:

- **Energy Optimization:**
  Trained employees can monitor and optimize energy consumption in real time, reducing operational emissions.
- **Efficient Waste Management:**
  Workers can implement best practices in waste management, supported by AI systems that minimize waste generation.

Conclusion: The Workforce of the Future in the Oil Industry

The implementation of AI in the oil industry not only transforms operations but also redefines the workforce's role. Continuous training and development in AI, data analysis, and predictive maintenance are essential to ensure a successful transition to a more efficient, sustainable, and profitable operational model.

Companies investing in their employees' development will be better equipped to face technological challenges and position themselves as leaders in an ever-evolving global energy market. The key to success lies in the synergy between advanced technology and a highly skilled workforce, ensuring a sustainable competitive advantage.

Optimization and Transformation in the Oil Industry with AI: A Detailed Analysis

The integration of AI in the oil industry is driving profound transformation, impacting operational structures, workforce dynamics, sustainability, and resilience in the global market. This technological advance redefines strategies for efficiency, innovation, and sustainability, providing companies with opportunities to maximize profitability and maintain long-term competitiveness.

1. Workforce Optimization and Cost Reduction

AI enables the automation of operational tasks and the optimization of human resource allocation, significantly reducing both operational (OPEX) and capital (CAPEX) costs.

Automation of Routine and Operational Tasks:

- **Autonomous System Management:**
  AI allows complex systems such as ESP pumps and separators to operate autonomously with real-time monitoring, eliminating the need for constant supervision and freeing human resources.
- **Reduction of Indirect Costs:**
  Decreased reliance on field personnel reduces indirect costs, such as insurance, training, and travel expenses.

Predictive Maintenance and CAPEX Optimization:

- **Extending Equipment Lifespan:**
  Predictive maintenance using AI identifies potential failures before they occur, extending equipment lifespan by 15-20% and reducing the need for costly replacements.
- **Quantifiable CAPEX Reduction:**
  A company with an annual CAPEX of $50 million could save $7.5-10 million annually by reducing premature replacements.

Global Impact:
The combined reduction in OPEX and CAPEX can increase operational margins by 15-20%, improving profitability and financial sustainability.

2. Sustainability and Environmental Responsibility

Sustainability has become a strategic pillar in the industry, with AI playing a fundamental role in this domain.

Energy Optimization and Emission Reduction:

- **Energy Consumption Reduction:**
  AI optimizes energy use in critical operations like pumping and separation, reducing energy consumption by up to 20%.

- **CO$_2$ Emissions Reduction:**
  This energy saving translates into significant emission reductions, enabling companies to meet environmental regulations and improve their carbon footprint.

Leak Prevention and Waste Management:

- **Early Leak Detection:**
  Smart sensors integrated with AI detect micro-leaks in their early stages, preventing major spills that could have devastating environmental and economic consequences.
- **Optimized Waste Management:**
  Processes for separating and treating water and solids are optimized, reducing waste generation by 10-15%.

Competitive Advantage:

Companies implementing these technologies will not only comply with stricter regulations but also access financial incentives and enhance their reputation with environmentally conscious investors.

3. Greater Resilience in a Volatile Market

Volatility is a constant in the oil industry, and AI provides tools to manage it effectively.

Dynamic Operational Adjustments:

- **Real-Time Production Optimization:**
  AI automatically adjusts production based on market conditions, avoiding losses from overproduction and maximizing revenue during periods of high demand.

Intelligent Inventory Management:

- **Demand Prediction with 90% Accuracy:**
  AI analyzes historical patterns and external factors to anticipate demand changes, enabling better inventory and logistics planning.

Impact on Results:

Companies using AI for production and inventory management can improve revenues by 5-10% annually, even in highly volatile markets.

4. Development of Specialized Talent and Knowledge Retention

The adoption of AI redefines required competencies, creating new technical and strategic roles.

New Roles and Competencies:

- **Automation Engineers and Data Analysts:**
  Workers need to master advanced technologies to manage automated systems and analyze large data volumes.
- **Advanced Maintenance Technicians:**
  These roles are essential for interpreting AI diagnostics and executing predictive maintenance tasks.

Continuous Development and Knowledge Retention:

- **Retention of Technical Knowledge:**
  Companies must ensure that the knowledge acquired in these areas remains within the organization by implementing professional development programs and innovative work environments.

Strategic Impact:

Creating a highly skilled workforce ensures a sustainable competitive advantage, positioning companies as leaders in innovation and efficiency.

5. Driving Innovation and Continuous Improvement

AI is not just an operational tool; it is a driver of innovation.

Development of New Technologies:

- **Insights from Operational Data:**
  Data generated by AI systems provides valuable insights for developing new technologies, such as autonomous drilling systems and advanced enhanced recovery methods.
- **Human-Machine Collaboration:**
  The combination of AI recommendations and human expertise fosters a continuous improvement environment where strategic decisions are data-driven.

Competitive Advantage:

Companies prioritizing innovation will be better positioned to lead in a market where efficiency and adaptability are crucial.

6. Long-Term Industry Impact

The adoption of AI represents a paradigm shift, ensuring sustainability and competitiveness in the long term.

Resilience and Adaptability in Complex Environments:

- **Preparation for Future Regulations:**
  By operating more efficiently and sustainably, companies will be better positioned to comply with stricter regulations.
- **Resilience Against Market Fluctuations:**
  The ability to quickly adapt to changes in demand and crude prices ensures companies can thrive even under adverse conditions.

AI implementation in the oil industry is not merely technological modernization; it is a strategic shift that transforms operational efficiency, sustainability, and resilience in a competitive global market. Companies embracing this technology will be better prepared to tackle future challenges, from market volatility to growing sustainability demands.

AI empowers human talent by freeing it from repetitive tasks, enabling a focus on strategic and innovative activities. This balance between advanced technology and a skilled workforce will be key to ensuring sustainable and competitive growth in the coming decades, paving the way for a more responsible and resilient oil industry.

Challenges and Future of AI in the Oil Industry: A Technical and Strategic Analysis

The implementation of artificial intelligence (AI) in the oil industry represents a profound transformation, redefining technical operations as well as the management of human and technological resources. However, this advancement introduces critical challenges that must be addressed with well-founded strategies in three key areas: workforce training, cybersecurity, and human-AI collaboration. Addressing these challenges effectively will not only amplify AI's benefits but also ensure its long-term sustainability and security.

1. Workforce Training and Retraining: Ensuring Technical Competence

a. Technical Context and Training Needs

The transition to an AI-driven environment requires a significant restructuring of workforce skills. This includes not only mastering technological tools but also interpreting and applying insights derived from operational data.

Key Technical Training Components:

- **Programming Languages and Analytical Tools:**
  Training in Python, R, MATLAB, and other critical languages for data manipulation and custom algorithm development.
- **Predictive Modeling and Machine Learning:**
  Instruction in supervised and unsupervised learning algorithms, frameworks like TensorFlow, PyTorch, and Scikit-learn, and techniques such as clustering, regression, and classification.
- **Data Integration and Systems Architecture:**
  Advanced knowledge of distributed systems like Apache Spark and Hadoop, essential for real-time data processing and scalable infrastructure.

b. Practical and Progressive Training

The approach should be hands-on, with simulations based on real operational scenarios. This ensures that personnel not only understand the theory but can also apply these skills in critical situations.

Quantifiable Benefit:
A structured training program can reduce learning time by 30-40%, accelerating AI integration into operations and improving overall efficiency within a year.

2. Cybersecurity: Protecting Critical Digital Infrastructures

a. Cyber Risks in an Automated Environment

Adopting AI increases the attack surface in operational systems, exposing companies to a variety of cyber risks.

Types of Threats:

- **Distributed Denial of Service (DDoS) Attacks:**
  These can paralyze critical systems like SCADA and DCS, affecting real-time process control and monitoring.
- **Manipulation of Operational Data:**
  Alterations to key data could lead to incorrect decisions, such as pressure adjustments that compromise operational safety.
- **Theft of Intellectual Property and Sensitive Data:**
  AI algorithms, predictive models, and production data are valuable assets that could be exploited by competitors or malicious actors.

b. Technical Cybersecurity Strategies

- **AI-Based Security:**
  Intrusion detection and prevention systems (IDS/IPS) powered by AI can identify anomalous patterns with 90% accuracy.

- **Network Segmentation:**
  Isolating key operational systems in segmented networks minimizes impact in case of an attack.
- **Data Encryption and Multi-Factor Authentication (MFA):**
  Ensures that only authorized personnel can access sensitive systems.

Quantifiable Impact:

Implementing these measures can reduce the likelihood of successful attacks by 60-70%, preventing estimated losses of $5-10 million per incident.

3. Human-AI Collaboration: Effective Integration of Technology and Human Judgment

a. Importance of Human-AI Interaction

AI excels at repetitive tasks and complex analyses, but human judgment remains indispensable in high-uncertainty situations.

Key Components:

- **Explainable AI (XAI):**
  Provides transparency in AI decision-making, allowing human operators to validate and adjust recommendations.
- **Intuitive Human-Machine Interfaces:**
  Customizable dashboards that facilitate AI interaction, offering real-time insights and enabling rapid manual adjustments.

b. Decision-Making in Complex Environments

- **Automation of Routine Decisions:**
  AI can handle predictable tasks with 95% accuracy, freeing up time for more critical decisions.

- **Human Oversight in Anomalies:**
  Events like unexpected pressure fluctuations require human intervention to interpret context beyond the data.

Operational Benefit:

Effective human-AI collaboration can reduce operational errors by 30-40% and shorten response times in critical situations to mere seconds.

4. Future Impact: Continuous Innovation and Operational Resilience

a. Data-Driven Innovation

AI not only optimizes current operations but also generates insights for developing new technologies, such as autonomous drilling and advanced enhanced recovery methods.

b. Resilience and Adaptability

The ability to quickly adjust to market conditions and regulations ensures that companies remain relevant and competitive in the long term.

Preparing the Industry for the Future

Integrating AI into the oil industry presents complex challenges requiring a comprehensive strategy. Advanced technical training, robust cybersecurity, and effective human-AI collaboration are not just immediate needs but foundational pillars for long-term success.

By addressing these challenges with technical precision, companies can maximize operational benefits and build a resilient, innovative, and secure environment, prepared to lead in an ever-evolving energy market.

Conclusion: Preparing the Oil Industry for an Intelligent and Resilient Future

The implementation of AI in the oil industry marks the beginning of an unprecedented strategic and operational transformation. While the benefits are significant—from cost optimization to improved sustainability and operational efficiency—the true key to success lies in how companies address the challenges associated with this technological transition.

1. **Mastery of Technology Through Continuous Training:**
   Adopting AI is not merely a technological shift but a rethinking of required skills and competencies. Companies investing in technical training will not only accelerate AI integration but also foster an environment of continuous innovation. Practical training in analytical tools, machine learning, and data management will be essential to ensure the workforce thrives in this new era.
2. **Cybersecurity: A Fundamental Pillar:**
   With increasing digitalization, protecting critical infrastructure and operational data becomes a priority. Advanced cybersecurity strategies, such as AI-based detection systems and network segmentation, will be crucial to mitigate risks and safeguard valuable assets.
3. **Human-AI Synergy: Harnessing the Best of Both Worlds:**
   While AI can process vast data sets and make operational decisions with speed and accuracy, human judgment remains essential for complex and uncertain scenarios. Companies that develop explainable AI (XAI) systems and intuitive human-machine interfaces will enable effective collaboration, ensuring more informed decisions and minimizing operational errors.
4. **Innovation and Resilience as Vectors of Competitiveness:**
   AI's ability to generate valuable insights and support technological innovation will be a key differentiator in a highly competitive and volatile market. Companies leveraging this data will not only optimize current operations but also develop new strategies and technologies, positioning themselves as industry leaders.
5. **Building a Sustainable Future:**
   AI integration enhances operational efficiency while contributing to a more sustainable

model. Reducing emissions, preventing spills, and optimizing energy use position companies as sustainability leaders, a critical factor for regulators, investors, and society.

Long-Term Vision

AI is not just a tool for immediate efficiency improvements; it is a platform for building a more resilient, innovative, and sustainable oil industry. Companies leading this transformation will be better prepared for future challenges, from market volatility to stricter environmental regulations, ensuring their relevance and competitiveness in an ever-evolving energy landscape.

Ultimately, the success of AI implementation will depend on a strategic balance between advanced technology, robust security, and a highly skilled workforce. This holistic approach will enable the oil industry to not only adapt but thrive in a world where technology and sustainability are the cornerstones of long-term growth.

Definitive Conclusion: Leading the Transformation of the Oil Industry with AI

The implementation of artificial intelligence (AI) in oil fields is not merely a technological enhancement; it is a revolution that redefines the industry from its foundations. AI enables optimization in every aspect of operations, from real-time monitoring to predictive maintenance management, achieving unprecedented operational efficiency. This advancement not only reduces costs and maximizes production but also meets global sustainability demands, significantly lowering the environmental footprint.

However, the true potential of AI lies in the collaboration between humans and machines. Far from eliminating the need for human talent, AI transforms traditional roles into highly technical and strategic positions. Engineers, data analysts, and supervisors trained in AI become the core of a rapidly evolving industry, ensuring that critical decisions are made with a balance of automated analysis and human judgment.

As companies adopt these technologies, they must address key challenges: continuous workforce training, robust protection against cyber threats, and the creation of infrastructure that enables efficient interaction between humans and intelligent systems. Organizations investing in these areas will not only adapt to a competitive and regulated market but also lead the transition toward a more agile and sustainable business model.

Ultimately, AI is not just a tool for improving profitability; it is a catalyst for the comprehensive transformation of the industry. Companies embracing this change with vision and commitment will position themselves as leaders in an era where technology and sustainability are fundamental to success. This is the future of the oil industry: intelligent, efficient, sustainable, and, above all, human at its core.

Definitive Conclusion: The Comprehensive Revolution of the Oil Industry with AI

The implementation of artificial intelligence (AI) in the oil industry is not just a technological breakthrough but a comprehensive transformation redefining every aspect of the sector. From operational optimization to sustainability, AI offers an unprecedented opportunity to enhance efficiency, reduce costs, and minimize environmental impact, ensuring competitiveness in an ever-evolving global market.

1. Comprehensive Operational Optimization

AI enables real-time monitoring and predictive maintenance that not only maximize production but also extend the lifespan of critical equipment such as ESP pumps and separation systems. This translates into a 15-25% reduction in OPEX and a total operational efficiency increase of up to 20%. Additionally, the ability to automate decisions in milliseconds ensures stable and productive operations even under dynamic conditions.

2. Sustainability and Resilience

The industry faces increasing pressure to operate sustainably. AI directly contributes to this goal by optimizing energy consumption and reducing CO_2 emissions by 10-20%, even integrating renewable energy sources into operations. Early leak detection and efficient waste management further reinforce environmental commitment, enhancing corporate reputation and meeting stringent regulations.

3. Transformation of Human Talent

Contrary to the perception that AI will replace human talent, this technology elevates the importance of technical and strategic roles. Automation engineers, data analysts, and advanced maintenance technicians become essential for interpreting complex data, validating automated decisions, and ensuring optimal and safe operations.

Human-AI collaboration not only minimizes operational errors but also maximizes response times in critical situations, improving decision-making by 30-40%. This balance between artificial intelligence and human judgment will be crucial for maintaining competitiveness and adaptability in the future.

4. Critical Challenges: Cybersecurity and Continuous Training

Digitalization and automation increase the attack surface for cybercriminals. Protecting critical infrastructures requires robust cybersecurity strategies, such as network segmentation, multi-factor authentication, and AI-based intrusion detection systems, capable of reducing the risk of successful attacks by 60-70%.

Simultaneously, continuous workforce training is indispensable. Companies investing in advanced training programs in programming languages, machine learning, and distributed infrastructure management will benefit from a highly competent workforce capable of efficiently integrating these technologies.

5. Innovation and Adaptability in Volatile Markets

AI's ability to anticipate demand fluctuations and adjust production in real-time ensures companies can operate profitably even in volatile markets. This adaptability allows revenue optimization of up to 10% annually, securing a strong position against geopolitical and economic changes.

Organizations adopting a proactive approach to data-driven innovation will not only be prepared for the present but will also lead the development of new technologies defining the industry's future.

6. The Future of the Industry: Intelligent, Sustainable, and Human

AI is more than an operational tool; it is a catalyst for the strategic transformation of the oil industry. Companies integrating AI into their operations will not only reduce costs and improve sustainability but also create a more agile, resilient, and future-ready business model.

In an environment where technological efficiency and environmental responsibility are essential, the true competitive advantage will lie in the synergy between advanced technology and skilled human talent. This balance will ensure that the oil industry not only thrives but also leads in a world increasingly focused on sustainability and innovation.

7. A Future Transformed by Artificial Intelligence

The revolution driven by artificial intelligence (AI) extends beyond optimizing individual processes within the oil industry; it reaches toward a complete redefinition of operational, economic, and sustainable models. In a global environment marked by volatility, environmental pressure, and rapid technological evolution, AI stands as a fundamental pillar for the transformation of this critical industry.

8. Operational Efficiency as the New Standard

AI has proven its capability to enhance operational efficiency at all levels, from automating repetitive tasks to making complex, data-driven decisions. This level of optimization not only reduces operational and capital costs but also ensures the continuity and stability of operations under challenging conditions.

9. Sustainability and Global Responsibility

In a global context where sustainability has become imperative, AI enables oil companies to significantly reduce their environmental footprint. From energy optimization to spill prevention and efficient waste management, these technologies facilitate compliance with strict regulations and strengthen social acceptance to operate.

10. Workforce Transformation

The industry's future does not discard human talent; it amplifies it. AI transforms traditional roles, creating a demand for advanced technical competencies in data analysis, automation, and cybersecurity. The workforce of the future will be more specialized, strategic, and crucial for maximizing the potential of these technologies.

11. Adaptability and Resilience in a Volatile World

AI's ability to predict and adapt to changing conditions positions companies as resilient leaders in a market characterized by price fluctuations, shifts in demand, and geopolitical challenges. This flexibility not only ensures survival in difficult times but also allows capitalizing on emerging opportunities.

12. A Call for Continuous Innovation

Implementing AI is not a destination but a pathway toward continuous improvement. Companies adopting an innovation mindset, supported by insights generated through AI, will be better positioned to develop disruptive technologies and lead the next wave of advancements in the industry.

13. Global Vision: Leadership in a New Paradigm

AI is transforming the oil industry into a more intelligent, sustainable, and human operation. Companies that understand this potential and implement it strategically will not only optimize operations and enhance profitability but also play a crucial role in shaping a more responsible and adaptable energy sector.

This is the challenge and opportunity of our time: to lead with technology, innovate with purpose, and build a future where the oil industry can thrive in harmony with a constantly changing world.

Key References for the Implementation of Artificial Intelligence in the Oil Industry

1. **Misagh, N., & Ashouri, M. (2016).**
   Spatial Modeling of Oil Exploration Areas Using Neural Networks and ANFIS in GIS.
   Available at: arxiv.org

2. **Li, J. X., et al. (2024).**
   Artificial General Intelligence (AGI) for the Oil and Gas Industry: A Review.
   Available at: arxiv.org

3. **Nordloh, V. A., Roubícková, A., & Brown, N. (2020).**
   Machine Learning for Gas and Oil Exploration.
   Available at: arxiv.org

4. **Smith, R. G., Baker, J. D., & Young, R. L. (1980s).**
   Dipmeter Advisor: The First Expert System for Oil Well Log Interpretation.
   Available at: Wikipedia

5. **Ortiz, A. (2023).**
   The Role of Artificial Intelligence in the Oil Industry.
   Available at: innovamas.nakasawaresources.com

6. **Rivera Torres, E., & Ceballos, N. (2022).**
   Advances in Digitalization in the Hydrocarbons Sector.
   Available at: palermo.edu

7. **Rovirosa Martínez, A. (2020).**
   Artificial Intelligence: A Necessary Investment for the Oil Industry.
   Available at: globalenergy.mx

8. **Villegas Morphy, L. (2023).**
   Artificial Intelligence for Decision-Making in Predictive Maintenance of Turbo Compressor Equipment in the Oil Industry.
   Available at: researchgate.net

9. **Martínez Bernardino, V. (2024).**
   The Age of Artificial Intelligence in the Oil Industry: Key Opportunities and Challenges

for the Sector in Mexico.

Available at: itpe.mx

10. **Authors of Eadic (2024).**

Artificial Intelligence Applied in the Oil Industry.

Available at: eadic.com

Glossary of Technical Terms

1. **Artificial Intelligence (AI):**
   A branch of computer science focused on creating systems capable of performing tasks that typically require human intelligence, such as pattern recognition, decision-making, and continuous learning. In the oil industry, AI is used to optimize operational processes, predict failures, and enhance overall efficiency.

2. **Machine Learning (ML):**
   A subset of AI focused on developing algorithms that learn from data and improve performance over time without explicit programming. It is crucial in the industry for reservoir behavior modeling, production optimization, and logistics improvement.

3. **Predictive Algorithms:**
   Mathematical procedures designed to analyze historical and real-time data to forecast future outcomes. In the oil industry, these algorithms are essential for anticipating pressure fluctuations, equipment failures, and market demand changes.

4. **Artificial Neural Networks (ANN):**
   Computational systems inspired by the human brain, consisting of interconnected layers of nodes that process input data to identify complex patterns. They are used for advanced analyses, such as predicting well productivity and optimizing separators.

5. **Predictive Maintenance:**
   A technique that employs sensors, AI algorithms, and historical data to predict when a component is likely to fail. This enables preemptive repairs, reducing unplanned downtime and extending equipment lifespan.

6. **Distributed Control System (DCS):**
   A technological platform used in industrial automation to monitor and control multiple processes across different locations. Integrated with AI, DCS enhances efficiency by automatically adjusting operational parameters in real-time.

7. **SCADA (Supervisory Control and Data Acquisition):**
   A combination of software and hardware systems that monitor and control critical infrastructure. In oil fields, SCADA enables remote supervision of wells, separators, and

other essential equipment, enhancing safety and reducing operational costs through data-driven decisions.

8. **Three-Phase Separators:**

   Equipment that separates a mixture of oil, gas, and water into individual components. AI optimizes their operation by automatically adjusting variables such as pressure and temperature, ensuring efficient and high-quality separation.

9. **ESP Pumps (Electrical Submersible Pumps):**

   An artificial lift system that uses a multistage submerged pump to increase crude oil production in low-pressure wells. AI enhances their operational efficiency by adjusting parameters such as motor speed and predicting failures to avoid interruptions.

10. **Cybersecurity:**

    Practices and technologies designed to protect critical systems, networks, and data from cyberattacks. In the oil industry, cybersecurity is vital to prevent unauthorized access that could disrupt operations or compromise confidential data.

11. **Network Segmentation:**

    A security technique that divides a network into smaller, controlled segments to limit the spread of potential attacks. In critical systems like SCADA and DCS, this ensures that a breach in one segment does not affect the entire operation.

12. **Explainable AI (XAI):**

    A field of AI focused on making machine learning models transparent and understandable to humans. This is crucial in the oil industry for validating automated decisions, especially in critical operations.

13. **Carbon Footprint:**

    The total measure of greenhouse gas emissions, such as carbon dioxide (CO_2), produced directly or indirectly by an organization's activities. AI-driven energy optimization helps reduce the carbon footprint in oil operations.

14. **Predictive Analytics:**

    The use of data mining, statistics, and machine learning techniques to forecast future events. In the oil industry, it is employed to anticipate demand, adjust production levels, and predict reservoir performance.

15. **Energy Optimization:**
    The process of adjusting equipment energy consumption to maximize operational efficiency and minimize waste. AI facilitates this optimization by dynamically adjusting configurations for equipment such as pumps and separators.

16. **Intellectual Property:**
    Legal rights protecting innovations such as AI algorithms, predictive models, and proprietary operational data. Safeguarding these assets is vital to maintaining a competitive edge in the industry.

17. **Intelligent Redundancy:**
    A feature of AI systems that ensures operational continuity by automatically managing multiple control pathways. This guarantees operations continue even if one part of the system fails.

18. **Data Lake:**
    A centralized repository storing large volumes of data in its raw format. In the oil industry, data lakes are used to consolidate operational data, facilitating advanced analytics and predictive model development.

19. **Integrated Systems:**
    Technological infrastructures that combine various systems, such as SCADA, DCS, and AI platforms, to work cohesively and efficiently. This allows for more comprehensive monitoring and control of operations.

20. **Operational Adaptability:**
    The ability of systems and equipment to adjust their operations in response to changing environmental conditions, such as pressure fluctuations or market demand. AI enables real-time adaptability, ensuring consistent and efficient production.

These references and terms underscore the transformative role of AI in the oil industry, from operational efficiency to strategic innovation, guiding the sector towards a more sustainable and competitive future.

Author's Postscript: The Time of Transformation

Throughout human history, never before have we witnessed a period of such rapid and transformative advancements as in the past 100 years. From the industrial revolution to the digital age, each discovery has laid the foundation for the next, bringing us to a present where artificial intelligence (AI) is emerging as the driving force behind the next great revolution.

However, the real question is not whether AI will fully integrate into our lives, but when and how it will do so in a more direct and visible way. Today, AI operates largely behind the scenes: optimizing internet searches, personalizing recommendations, and enhancing industrial efficiency. Yet, a future where this technology becomes a visible, conscious, and everyday part of our existence is no longer a distant dream. The question is: are we ready?

How Soon Will We Live This Future?

Analyzing the exponential rate of technological innovation, it is foreseeable that in the next 10 to 20 years, AI will become an integral and visible presence in all aspects of human life. This will not be a subtle change but a quantum leap into a world where:

- Machines will make critical decisions in fractions of a second.
- AI systems will collaborate closely with us in real time, not just as assistants but as equal partners.
- Physical and cognitive work will be optimized to the point of redefining our understanding of employment and productivity.

The mystique of this advancement lies in the fact that we are venturing into uncharted territory. Unlike any previous revolution, this one will not only change how we work but also how we think, learn, and live.

The Jobs of the Future: Adapting to a Redefined World

In this new paradigm, traditional roles will disappear or evolve drastically. New professions will emerge, some of which we cannot yet fully imagine. However, some key roles are already on the horizon:

1. **Artificial Consciousness Engineers**
   They will not only design intelligent systems but also work on their capacity to make ethical decisions aligned with human values.
2. **Neurotechnology and Human Connection Specialists**
   As AI integrates more deeply into our bodies and minds, these experts will ensure a safe and efficient symbiosis.
3. **Digital Philosophers and Ethicists**
   The advancement of AI poses profound ethical dilemmas. What does it mean to be human in a world where machines possess capabilities equal to or greater than our own? These specialists will be vital in guiding our collective decisions.
4. **Mixed Reality Designers**
   Responsible for creating environments where the line between virtual and physical blurs, offering new ways to experience life.
5. **Educators in Artificial Intelligence and Humanism**
   In an increasingly automated world, it will be essential to teach new generations not only technical skills but also ethical values and social abilities that distinguish humans from machines.

Preparing Future Generations: The Challenge of Educating for the Unknown

How do we educate future generations to thrive in a world we do not fully understand ourselves? The key lies in fostering adaptability, creativity, and critical thinking. These skills will be essential to navigating a future where the only constant is change.

- **Adaptability:** Jobs and roles of tomorrow will be fluid. We must prepare young people for continuous learning, capable of retraining quickly as needed.
- **Creativity:** In a world where AI handles data and optimization, the human capacity to imagine new possibilities will be invaluable.

- **Critical Thinking:** With the proliferation of information generated by AI, the ability to discern what is true, ethical, and useful will be vital for decision-making.

What Does the Future Hold?

The current pace of technological advancement suggests that by the mid-21st century, humanity will be living a reality that just decades ago seemed like science fiction. The Technological Singularity, a theoretical point where machine intelligence surpasses human intelligence and becomes self-sufficient, could be reached in the next 30 to 40 years.

This moment could redefine the very essence of our existence. Will we be passive spectators or active protagonists in this transformation? Humanity's destiny will depend on how we choose to use this technology and our ability to integrate progress with the preservation of our humanity.

The Call of the Future: Reflection and Preparation

The future is no longer a distant horizon; it is rapidly unfolding before us. This is a call to all visionaries, educators, leaders, and citizens: we must embrace uncertainty with courage, prepare with wisdom, and act with responsibility. The time of transformation is near, and the decisions we make today will be the legacy for future generations.

www.ingramcontent.com/pod-product-compliance
Lightning Source LLC
Chambersburg PA
CBHW050254220526
45465CB00002B/675